HEINKEL He 100

World Record and Propaganda Aircraft

Hans-Peter Dabrowski

Left: Photo montage: In the foreground, an He 100 D. In the background, the Rechlin aircraft test center with the west front of the hall, and the Müritz (lake) behind it.

SCHIFFER MILITARY HISTORY
West Chester, PA

Cover artwork by Steve Ferguson, Colorado Springs, Colorado.

PHOTO CREDITS:
Heinkel archives, Stuttgart,
and the Dressel, Griehl, Hartwig, Koos, Niedree, Nowarra, Petrick and Schnittke collections.

SOURCES:
Heinkel archives, Stuttgart,
Drawings and correspondence: Beist, Beauvais, Dieterle, Kerle, Köhler, Lange, Matthaes, Meier, Menk, Schnittke.

SELECT BIBLIOGRAPHY:
Heinkel, *Ein stürmisches Leben*; Köhler, *Ernst Heinkel, Pionier der Schnellflugzeuge*; Lange, *Typenhandbuch der deutschen Luftfahrttechnik*; Nowarra, *Die deutsche Luftrüstung*.

Translated from the German by Dr. Ed Force.

Library of Congress Catalog Number: 91-66339.

Printed in the United States of America.
ISBN: 0-88740-345-X

This title was originally published under the title,
Heinkel He 100, Band 120,
by Podzun-Pallas Verlag, Friedberg.

Published by Schiffer Publishing, Ltd.
1469 Morstein Road
West Chester, Pennsylvania 19380
Please write for a free catalog.
This book may be purchased from the publisher.
Please include $2.00 postage.
Try your bookstore first.

Above, a Messerschmitt Bf 109, below a Heinkel He 112. The comparison flying of the two types was won by the Bf 109, which became the Luftwaffe's new standard-bearer.

FOREWORD

The Heinkel He 100 single-seat fighter was built a good fifty years ago, and there are few eyewitnesses alive today who can report on it; many details have been forgotten, and a large part of the Heinkel works was destroyed when the war ended.

The special role of the He 100 as a major figure in the bewildering whirl of Third-Reich propaganda is another reason why it is scarcely possible to reconstruct the full history of this interesting aircraft today, and the following presentation, despite all the reliable sources, -can be regarded only as a fragment.

In 1938 and 1939, numerous reports on the records set by the Heinkel He 112U "fighter plane" appeared in newspapers and magazines, and beginning about 1940, periodicals such as "Der Adler", "Die Wehrmacht", "Der deutsche Sportflieger", "Luftwelt" and the like portrayed the He 113 fighter plane again and again. Almost all the books that concern the Luftwaffe in general have included at least one picture of the "new German fighter in action." All these reports have had one thing in common: they described an aircraft that actually did not exist, for the type number 113 was never entered in the official list of the Reich Air Ministry. It was always the He 100!

In many ways it was an unusual airplane; in 1939 it brought the absolute world speed record to Germany for the first time. Cooling, hydraulics, pumping system, riveting —all was new and different from before.

Originally planned as a fighter, it fulfilled Ernst Heinkel's ambitious wishes to set records, but instead of being built in large series and competing with the Messerschmitt Bf 109, only 23 planes were built. Some of them were sold to the Soviet Union and Japan, others were used for propaganda purposes or served as test models (for example, for nose-wheel tests for the He 280 jet, as crash cells, etc.). Some He 100's were set aside for the Heinkel factory protection unit.

Above: the Focke-Wulf Fw 187 as a twin-engine solution to the problem of building a superfast pursuit fighter; below: Heinkel's single-engine He 100 V-1 super pursuit fighter. Both models were far superior in speed to comparable types.

The Italian Francesco Agello held the absolute world speed record until 1939, having set it with this Macchi Castoldi MC 72. Heinkel set out to break it with the He 100 — and it was 709.21 kph at that! He utilized highly developed aerodynamics against the brute-force horsepower of the MC 72.

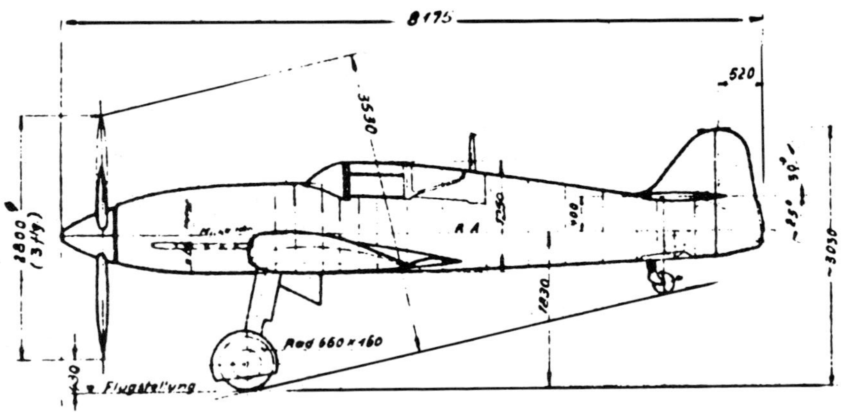

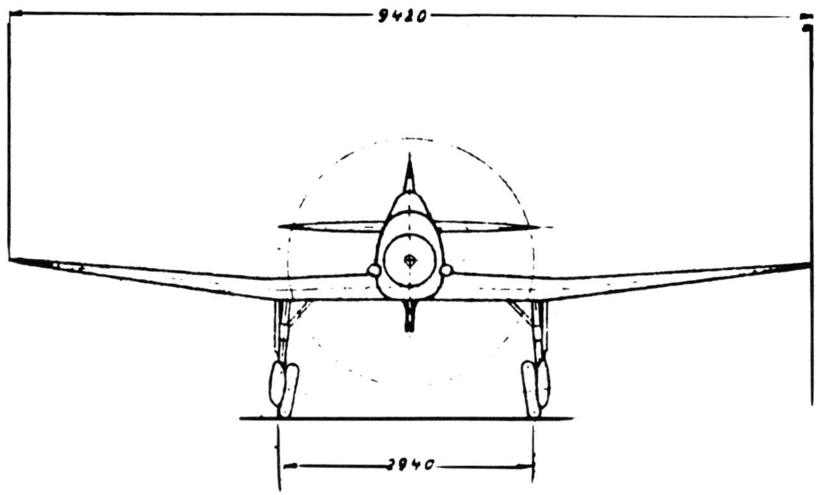

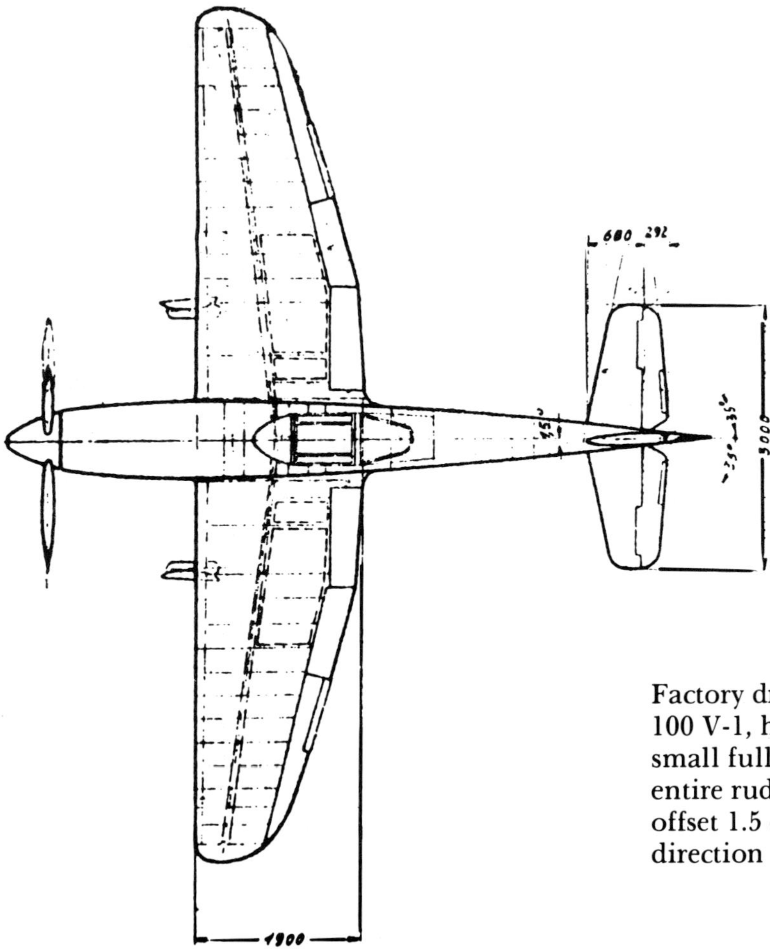

Factory drawing of the He 100 V-1, here still with a small full-length rudder, the entire rudder assembly is offset 1.5 degrees from the direction of flight.

DEVELOPMENT

When the Reich Air Ministry (RLM) issued specifications for a new standard fighter plane for the Luftwaffe in 1935, the following aircraft manufacturers offered the following models: Arado Ar 80, Focke-Wulf Fw 159, Heinkel He 112 and Messerschmitt Bf 109 (actually outside the competition).

Arado and Focke-Wulf soon dropped out, and Heinkel proved inferior to Messerschmitt in comparison flights in the spring of 1936. The reason why the He 112 could not beat the Bf 109 was that it was "not finished", as can be read in detail in Heinkel's biography, "Ein stürmisches Leben" (A Stormy Life). In addition, the He 112 was built in more complex form and was thus harder to maintain than the Bf 109.

This lack of success bothered Ernst Heinkel, for he was firmly convinced that he could build a better fighter plane than Messerschmitt. Then came the RLM's encouragement (through Fliegeroberstabs-Ing. Lucht) to build a superfast pursuit fighter — that was at the beginning of 1936.

The twin brothers Walter and Siegfried Günter ranked among Heinkel's best developmental engineers and had already been thinking of making further improvements to the He 112. Walter Günter realized that nothing more could be improved, and instead he designed a completely new airplane. The "Super-VJ" was given project number 1035, and preparations were made in strict secrecy to build mockups — Heinkel Memorandum No. 3657 of January 31, 1936: "The mockup in to be finished by us . . . as of the beginning of May . . . and ready to present to the RLM . . . and prior to that, no one at the RLM is to know of the existence of the mockup."

Left: The twin-engine Breda 88 destroyer with which the Italian pilot Niclot held the speed record for land planes.

The He 100 V-2 in which Ernst Udet brought the record for land planes to Germany on June 5, 1938, reaching 634.73 kph, more than 80 kph faster than Niclot. The plane's underside was painted yellow for the record run to stand out better against the sky.

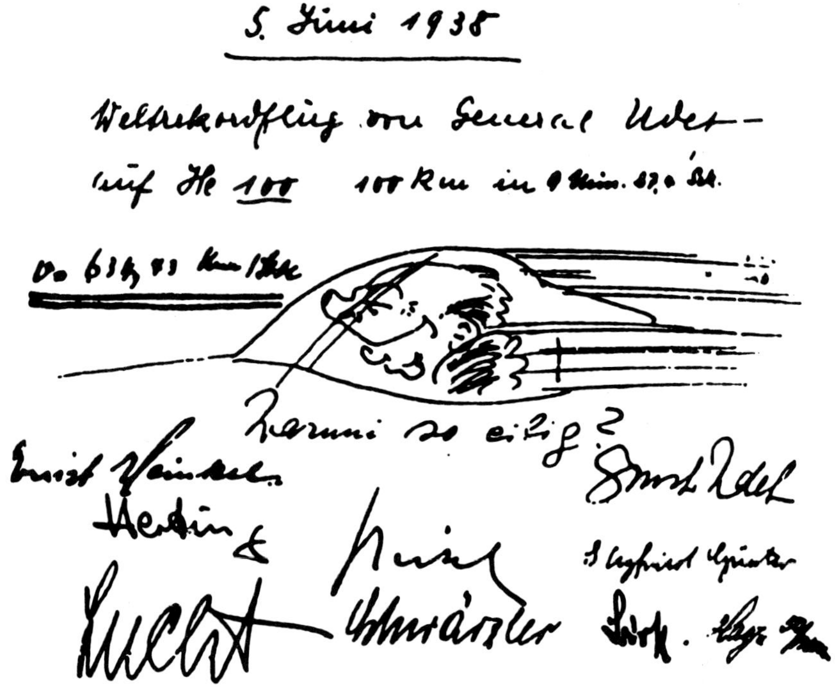

Udet in the plane, a cartoon drawn for the memorable occasion, the record plane and the people concerned: from left to right, Josef Köhler, Karl Schwärzler, Siegfried Günter, Ernst Udet, Ernst Heinkel, Heinrich Hertel, Fritjof Herting.

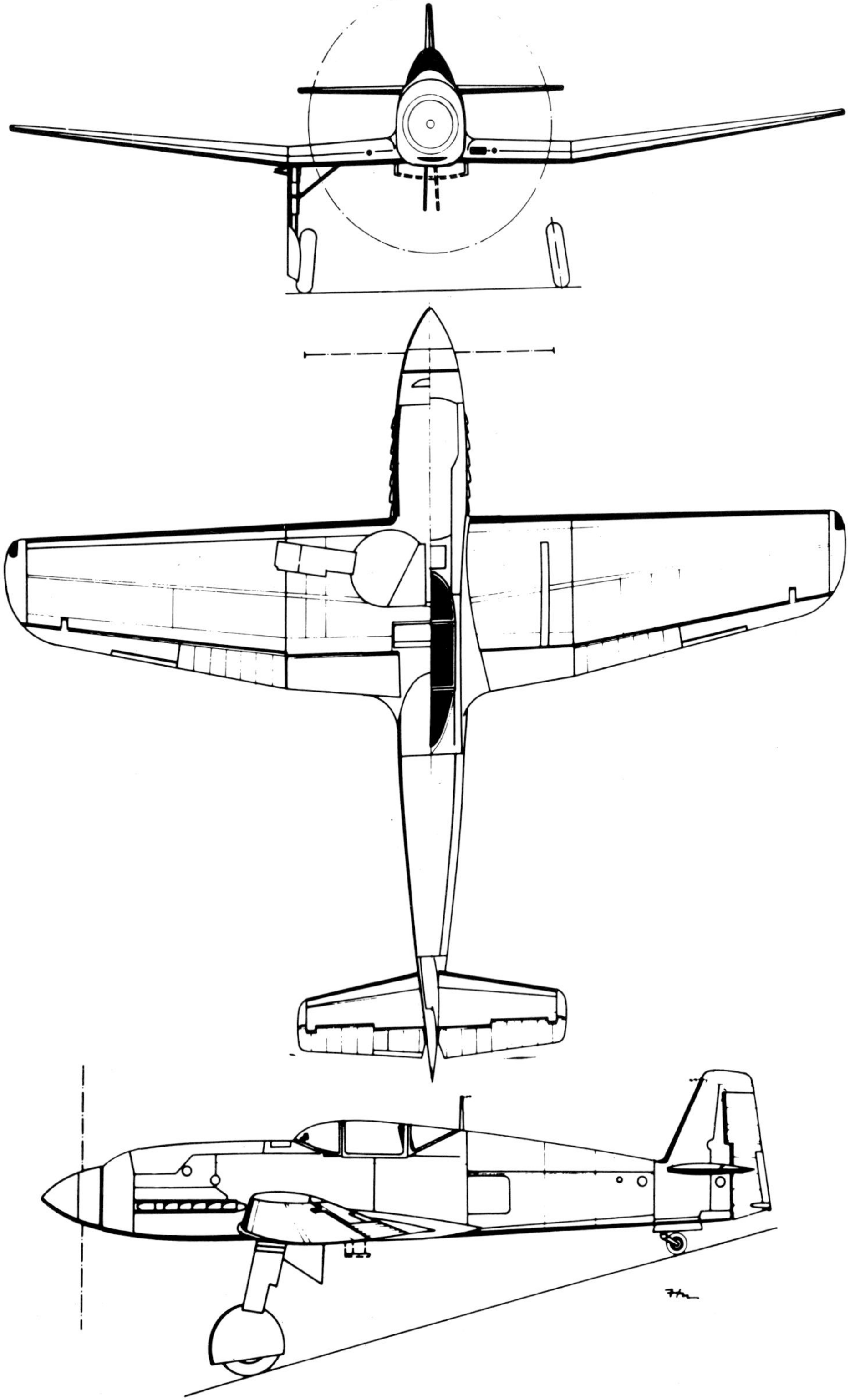

Left: He 100 V-4, the first model for the planned series.

Focke-Wulf also took up the idea of the super-pursuit plane and designed a twin-engine craft, the later Fw 187. This type was test-flown by Prof. Kurt Tank in the summer of 1937, but V-3 and V-4 were rebuilt into two-seaters while under construction. The few planes that were built were used for propaganda purposes just as was the He 100, as well as serving in Focke-Wulf's industrial protection unit, apart from short-term assignments in Norway and Germany. The people at Heinkel were well aware of this competitive pursuit-fighter development in 1936.

Walter Günter died in an auto accident on May 25, 1937, and his brother Siegfried took over the work on Project 1035. The task was clear: First a speed of 650 kph was to be striven for, and then came the direct comparison with the Bf 109: 5% less weight, smaller control surfaces, and some 70 to 80 kph higher speed. In the very first designs, the external form of which was scarcely changed, created a sleek, remarkably fine-looking craft; the high speed at which it was meant to fly could literally be seen.

While Heinkel worked on Project 1035 in the middle of 1937, taking up the problem of armament for the future fighter and wrestling with design difficulties, the 4th International Aircraft Meeting was taking place in Dubendorf, Switzerland, from July 23 to August 1, 1937, and Messerschmitt's Bf 109 was setting records there: Major Seidemann won the Alpine Flight with V-9, with Bf 109's also taking second and third places. Carl Francke won the race around the Alps and the alpine flight for military planes; they won special tests, climbing and diving competitions (Carl Francke was victorious here too), and the alpine flight for three-man crews — only Generalmajor Ernst Udet, Chief of the Technical Office of the RLM, had bad luck when his Bf 109 was badly damaged in a crash-landing caused by engine trouble.

The new Heinkel fighter, though, was supposed to be capable of beating Messerschmitt. It was "completely new" in every respect, including assembly: The "explosive rivet process" invented by the Butter brothers was used, and was planned on for use in series production for the first time. These rivets worked especially well in hard-to-reach places and components that were not under the greatest pressure. They were hollow inside and filled with a small explosive charge that was detonated electrically. The rivets were driven in by the explosion and were bomb-proof. Otto Butter and Fr. Scherer informed Ernst Heinkel of the following on October 27, 1937: After finishing the wings of P 1035 and after counting all parts precisely, these wings . . . show, compared to the He 112 . . , the following great superiority:

	He 112	P 1035
Individual parts	2,558	969
Rivets	26,864	11,534
Normal parts other than rivets	1,279	2,186

This shows a saving in the P 1035 of 62.1% in individual parts and 57.0% in rivets. The saving in rivets alone saved 1150 hours of work time for one pair of wings.

The explosive rivet process was indeed used for P 1035, but the RLM denied its value completely, refused it and sold the patent to the USA in 1940!

A further innovation was the surface cooling. The plane was powered by an in-line 12-cylinder Daimler-Benz DB 601 A (water-cooled) engine. The cooling water was under pressure and reached about 110 degrees celsius. The motor withstood these high temperatures. If the water expanded, it formed steam . . . Walter Günter developed this concept of surface cooling. In the Heinkel research department (directed by Dr. Matthaes), the specialists were Dipl.Ing. Jahn and Jahnke, Moser, and sometimes Kraeft and Siebenlist. In P 1035, the steam streaming out of the centrifugal steam separator was thus precipitated into cooling pockets that ran from the leading edges of the wings to their main spars. They were made of a copper-free aluminum alloy and welded electrically (later on, fuselage parts were added to the cooling surfaces). The condensed water was moved at first by membrane pumps, later by injectors. 22 small electric pumps drove the cooling water back to the engine. 22 indicator lights were mounted on the control panel in the cockpit, and one lit up when there was a disturbance in one of the pumps. To be sure, the surface cooling system was quite prone to breakdowns (A memo from Ernst Heinkel, dated March 31, 1938: "Kleinemeyer told me today that Schwärzler . . . told him . . . he thought it was only right that the cooling did not work and he was finally called in to work on it.")

Beyond that, the whole plane was built compactly, for example, the fuselage shell was extended to serve as an engine mount, the landing gear shafts were made very narrow

Above: He 100 V-3, the plane with which the world record was to be broken. A special propeller with short, very wide blades, built by VDM, was planned.

Right: Hans Dieterle preparing for a flight with V-3 at Rostock-Marienehe, where all the He 100's were built.

(hydraulic hoses were damaged), etc. The new cooling methods were meant to dispense completely with a belly radiator but include a retractable cooler only needed for take-offs.

All in all, though, this plane was capable of setting new records for Heinkel! What if a motor that would provide even more power had been installed? What if the exhaust gases had been used to provide more thrust? What if new airscrew designs and even greater weight reduction could have resulted in even higher speeds? All those things were considered by Heinkel . . . and then a man known to all had to make the record flights: the Chief of the Technical Office in the RLM, Generalmajor Ernst Udet, who was also a world-famous aviator!

At the end of October 1937, P 1035 was given a new name: He 100. Right after that, an offer was submitted to the RLM with detailed costs and dates for the delivery of the first three model planes to the test center. In a discussion at the Heinkel works on November 15, 1937, the He 100 V-1 was promised as ready to fly by January 15, 1938 (three days before, on November 11, 1937, Dr.-Ing. Hermann Wurster, in Messerschmitt Bf 109 V-13, set a record for land planes of 610.21 kph). Now Heinkel was determined to break all possible speed records and decided to build a V-4 as well, while V-1 was to be used for record flights. New wings had to be built for the purpose. If this work was finished before the planned arming of V-2, then the second plane was to be used for the record runs.

In the Heinkel conference of December 16, 1937, with Udet and (the meanwhile promoted) Chief Staff Engineer Lucht, not only was the armament of the He 100 decided on, but so was the following:

Above: The record flight of V-8, made by Hans Dieterle, won him the title of "Flugkapitän" — at 23, the youngest man to hold it.

Left: The timing station during He 100 V-8's record flight, which was flown between Neuruppin and Kremmen.

V-8 was not painted or numbered during the record flight. The dark spots are spots that were puttied and sanded.

After the record flight, V-8 was given a military paint job. The shortened wingspan of 7.60 meters is especially striking in this photo.

This special Heinkel publication described the record flight in detail and naturally referred to V-8 as He 112 U.

Other views of He 100 V-8. Small photo: with civilian registration. Large photo: here it can be seen clearly that the civilian registration, D-IDGH, was simply covered over — now the plane was a "fighter." The relatively small national emblem is explained by the fact that the front of the rudder served as a cooling surface and was not painted over (which was probably done here for propaganda purposes with paint that could be washed off: the waviness of the material pasted on came about in the same way). Normally, the correct registration would have been 42+C11, indicating a plane of Luftkreis Command IV in Münster.

In this test plane it can be seen clearly that the front panel of the cockpit canopy was curved (RLM requirement on account of the reflective sight for the weapons). The He 100 V-3 and V-8 record planes, though, had a flat windshield.

Detail photo of "series production of the He 113."

V-1: Testing, record flights, no weapons or FT installed. V-2: Used as V-1. V-3: Record plane with no military components and with a new windshield. V-4 was intended as the first model plane for the planned series with simple assembly and preparation. Udet was present and knew of the record preparations; he also knew that ernst Heinkel was not thinking of series production. Either he did not believe that the Heinkel plane could break the existing records, including the absolute world speed record then held by Italy, at least not with the amount of horsepower that could be produced then, or he wanted at least to make a record attempt (which had never been achieved at Dübendorf) and wanted to "wait and see" what could be done with the He 100 (which is more likely). In any case, this was no help to Heinkel's plans.

Udet made his record flight with He 100 V-2 on Whitsun of 1938, and certainly not by accident, as he seems to have been happy to admit, for this flight was prepared for carefully and gained a lot of prestige for Ernst Heinkel — so much that in March of 1939 a relatively unknown young pilot named Hans Dieterle was able to set a new absolute speed record with V-8.

The text that follows will provide as much information as possible on the individual test planes and the few D models. Since the planes were not always completed in numerical order, the chronological account does not agree with the order of the test-model series.

Upper left and below: Side panels of the engine mounts and engine installation: It is clear that the plane was made very narrow.

He 100 V-5 HE+BE being prepared for the reenacted record flight.

Left: At left Gerhard Nitschke, who had to give up V-3, at right Hans Dieterle, who set the world record with V-8.

He 100 V-1, factory number 1901

Registration D-ISVR. First flight January 22, 1938 at Rostock-Marienehe. Test flights at Rechlin as of March 1938. Surface cooling and additional retractable belly radiator, small rudder rounded at the top. higher elevators than the later models. Water-cooled DB601A engine. Until October 28, 1938 there were 192 points of objection to this plane, which shows what difficulties had to be surmounted with the new type.

After the non-aggression pact with the Soviet Union was signed (August 23, 1939), the plane was sold to that country.

He 100 V-2, factory number 1902

Registration D-IUOS. First flight May 17, 1938, with enlarged, likewise retractable belly radiator, side fins also larger than V-1. 100-kilometer record flight with normal DB production motor, flown by Ernst Udet on June 5, 1938 (Whitsun). He attained a speed of 634.73 kph thus breaking the record set by the Italian Niclot with the powerful twin-engine Breda 88 by a good 80 kph. Of course, before V-2 could be flown by Udet, some time passed, as numerous problems had to be solved. Francke of the Rechlin Test Center finally had no more objections to Udet's making the flight as of April 12, 1938, and Schwärzler, chief of Heinkel's design department, also ordered a seat cushion to be prepared for the short general. The Rechlin pilot Dipl.Ing Menk was also detailed to prepare V-2 for Udet at the Heinkel works. That took a good three weeks, and during this testing also flew some 30 kph above the official record, which he kept to himself for good reasons. The plane was always difficult: the landing speed was so high that at the Heinkel factory airfield, depending on pilot and conditions, the brakes sometimes had to be used up to the last 100 meters. The plane ran on a special fuel that eroded the metal parts; after every flight all the fuel lines, containers and motor had to be flushed out.

Left: The last test model, D-ITLR; note the cooling surfaces and the registration numbers moved far back on the fuselage.

Photos from the reenacted record flight made by HE+BE. The negatives, not cleared by the censors, were made unusable with an office punch.

Deplaning after the record flight: at left the real thing, at right the reenactment. The man in the leather jacket is Jupp Köhler, "Hein-Gummi", "maid-of-all-work" at Heinkel and responsible for special flights.

For the reenacted flight they did not go to the trouble of puttying over and sanding down rivets, joints and such.

Manager Karl Hayn, Prof. Ernst Heinkel and Director Robert Lusser stand before He 100 V-7.

Mechanic Wübbelmann, test pilot Dieterle and Köhler, the special flight director, before He 100 V-8.

The record flight almost did not take place. Because of the new type of surface cooling and the warning lights on the instrument panel, Udet was instructed to break off the flight at once if one of them lit up, but he did not hear the directions clearly and, quite bewildered, asked after the flight why those "silly" lights had lit up. It was very simple: Every burning light meant a failed electric pump; after a short time the engine would overheat, the oil would burn and the pistons would seize, but once again, everything went well.

On October 22, 1938 the tail wheel malfunctioned and the plane was damaged. It was scrapped on September 7, 1939.

He 100 V-3, factory number 1904

Registration D-ISVR (taken from V-1). After being rebuilt and equipped with short speed-record wings, registered as D-IDGH. After V-2, V-3 was intended exclusively for record runs. With it, Heinkel's chief pilot Gerhard Nitschke was supposed to break the record held to then by the Italian Francesco Agello and his Macchi Castoldi. The MC 72 set the world record of 709.21 kph on October 23, 1934. This speed was attained with a 3100-horsepower double engine, since the seaplane, equipped with its pontoons, encountered much air resistance. In He 100 V-3, on the other hand, all parts that created resistance were avoided. A rounded cockpit canopy, retractable landing gear and belly radiator, recessed rivets and the like served to improve aerodynamics. The oil was cooled as follows: The oil was pumped through a heat exchanger that was surrounded by methyl alcohol with a low vaporizing temperature. When the oil became hotter than 65 degrees Celsius, the alcohol evaporated, drawing warmth from the oil and thus cooling it. The alcohol vapors were drawn into the control surfaces, cooled off there, condensed back into liquid alcohol and were pumped back to the heat exchanger.

V-3 first flew in June of 1938; on one of Nitschke's flights in August of that year the retractable landing gear stuck in a half-extended position and, in agreement with the flight leadership, the pilot bailed out with his parachute. The plane crashed on the factory airfield at Rostock-Marienehe.

Above: A test model has landed. Note the white condenser on the fuselage, cooling surfaces on the rudder and oversize German cross far inboard on the wing.

Right: A Russian delegation with Hans Dieterle before the He 100 test model with a large cross on the underside of the wing.

The white letters —+NE, white cooling surfaces on the undersides of the wings, and opening for the motor cannon on the nose of the propeller boss can be seen on this He 100 D.

Ah HE 100 in the Soviet Union, at ZAGI (Zentralnyi Aerodinamiceskij i Gidrodinamiceskij Institut: Central Aerodynamic & Hydrodynamic Institute), an institution similar to the German AVA (Aerodynamische Versuchsanstalt: Aerodynamic Test Center) at Göttingen.

Below: Ge 100 GP+AI in its usual version, with surface cooling.

He 100 D shortly before take-off.

This He 100 D has normal cooling; the octane triangles are seen in the cross emblems.

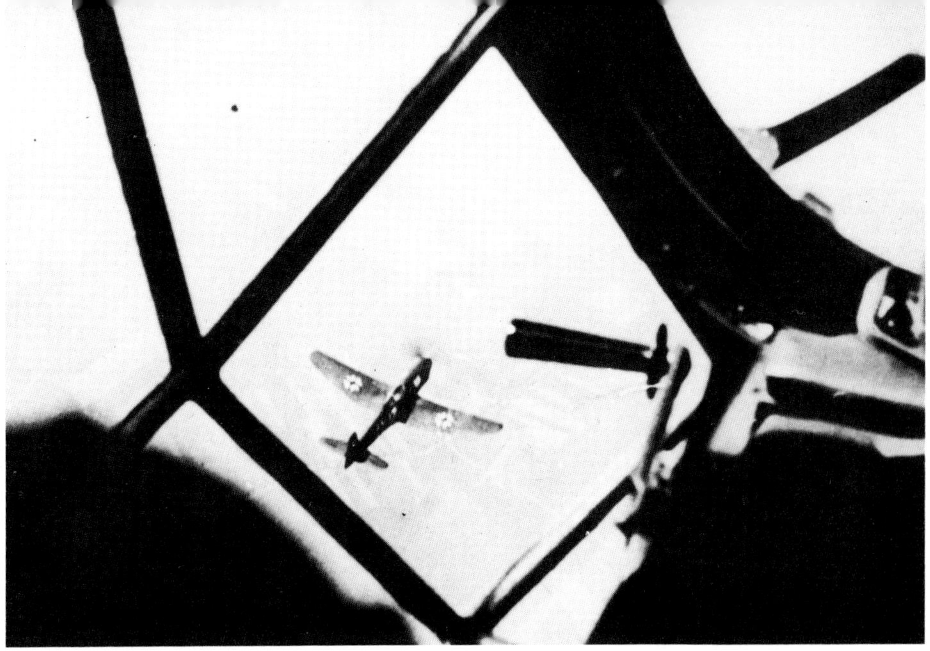

Various views of an He 100 D as "escort fighter."

On August 20, 1938 the French General Vuillemin and his retinue visited the Heinkel works at Oranienburg, and as he flew quite slowly in a Fieseler "Storch", factory pilot Dieterle thundered over the field with a He 100 and everything in it, and then landed. This scene was portrayed deliberately; the He 100 was "sold" to the general as a production plane, and the general was, in fact, very impressed.

He 100 V-4, factory number 1903

Registration D-IRCN, prototype of the planned series, simplified assembly, and the first model that was fully armed. Flight testing under high standards (factory pilot Ursinius attained about 850 kph on a straight-line flight). This plane's first flight took place in September of 1938; on October 22, 1938 the landing gear collapsed while standing and the plane was badly damaged. On March 16, 1939 the Rechlin engineer-pilot Heinz Beauvais flew B-4, early in November 1939 it became known that the He 100 was not to be built in series. The plane was later sold to Russia.

He 100 V-5, factory number 3001

Registration D-ISRO, first flight on November 16, 1938, flight testing at Rechlin. Straight-line flights were made by pilots Francke, Bader, Thoenes and others. Under registration HE+BE, Dieterle's record flight in V-8 was switched to V-5 so as not to damage the original plane. Since the Heinkel record did not last long before being broken by the Me 209, the Heinkel staff wanted revenge . . . Sold to Russia.

He 100 V-6, factory number 3002

Registration D-IACI, first flight February 1939, was used at Rechlin by Dipl.Ing. Baist for thoroughgoing engine testing as of May 17, 1939. The results were not completely satisfactory, as, for example, the cooling system still did not work problem-free, and the plane was built "too narrow", which made servicing difficult. Then on June 9, 1939 Baist had the all too common "experience" of landing-gear hydraulic failure. Because of a damaged hose, only one wheel was deployed, and the plane made a classic "circle dance" on the Rechlin field. Thanks to Baist's remarkable flying skill, the plane was only slightly damaged, and test flights could be resumed six days later. The flights ended at Rechlin on August 5, 1939, and the plane was later sold to Russia.

Right: Firing drill on the Baltic coast.

The ground crew at work.

From test plane to "fighter", complete with score stripes and mascot.

This is roughly how General Vuillemin experienced the He 100.

"Combat readiness" is demonstrated here.

A few last tweaks to the motor, then away with the brake blocks and off she goes over the Mecklenburg landscape.

Photos showing more than one He 100 in the air are rare.

Here it is a lightning flash; at other times it was an admiral's hat or a moon, passing for the emblems of units that did not exist.

The fast flyers in the air and on the ground. All the available models were once lined up in a row to give the impression that there were large quantities of the "He 113."

Several pictures were taken of this He 100D with the white number 12 in flight. They appeared again and again as "action photos" in periodicals and books from 1940 to about 1942.

This and the next page show additional pictures of the same themes. It can be seen that the few available planes were photographed from every conceivable angle at a few localities, and the photos were used for propaganda purposes.

21

Most in-flight pictures were taken from an He 111. On the wing of the plane on the ground a rod can be seen. It was painted red and appeared when the landing gear had been deployed without difficulty. There was often trouble with the landing gear, especially in the test models.

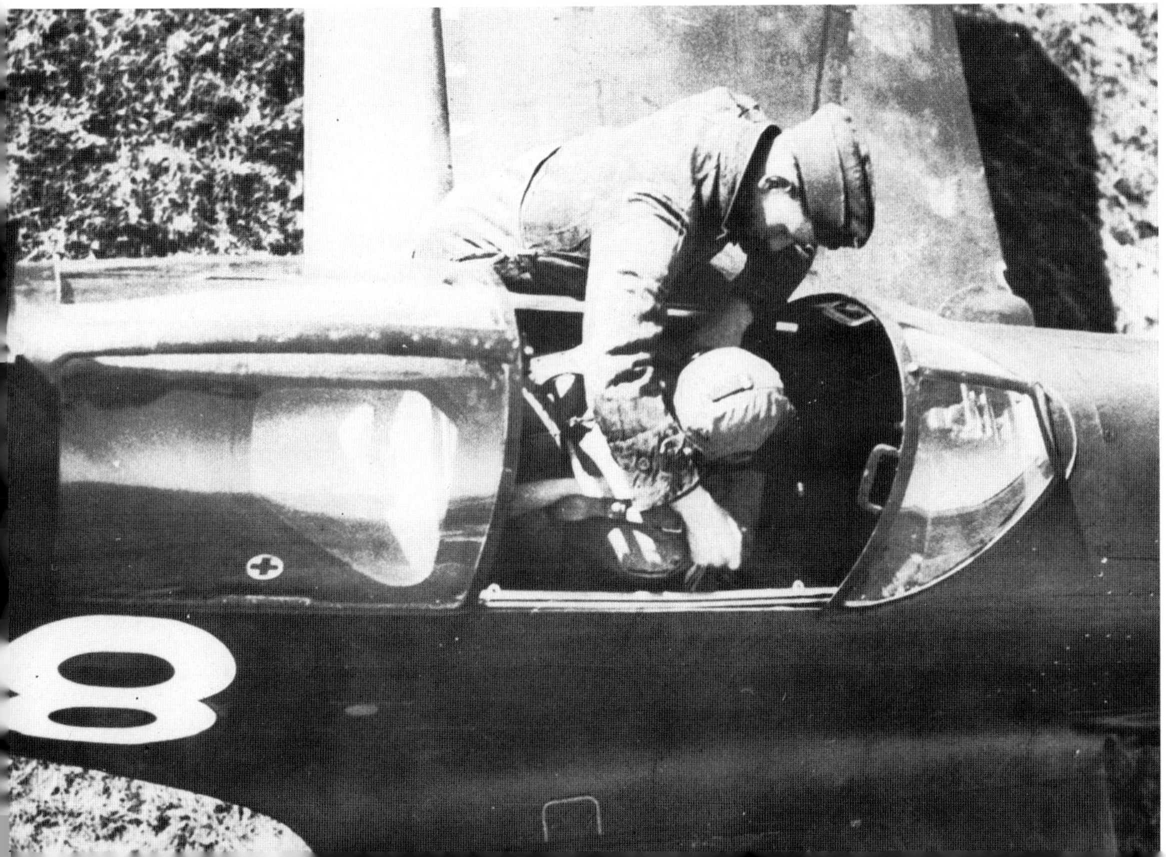

And again and again, the last few tweaks and preparations for take-off. Let it be noted here that the 100D always had the standard paint job, with segments of blackish green and dark green above and light blue below. Spot or dot camouflage was unknown. The first test models were covered with either gray paint or colorless corrosion protection.

As the "He 113 single-seat fighter" with a cannon firing through the propeller boss and machine guns in the wing roots — the British were obviously convinced they had encountered this type. Perhaps the factory protection unit actually saw combat?

Two examples of "combat reports" in German magazines: "Proved splendidly in front-line combat" and "air war over Norway", which no one has been able to authenticate to this day.

The He 100 also "saw service" as a "night fighter", as can be seen on this page and the next.

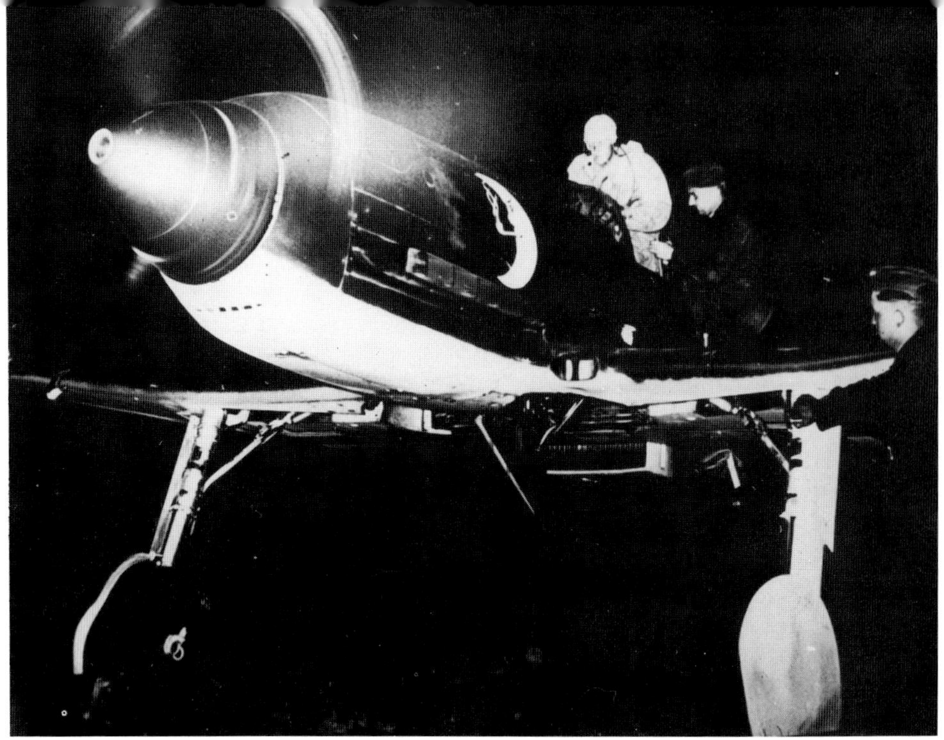

The He 100 V-10 was equipped with the speed-record wings and the flat spherical cockpit canopy of the V-8 and displayed in the German Museum at Munich as the He 112 U. The plane was destroyed in a bomb attack.

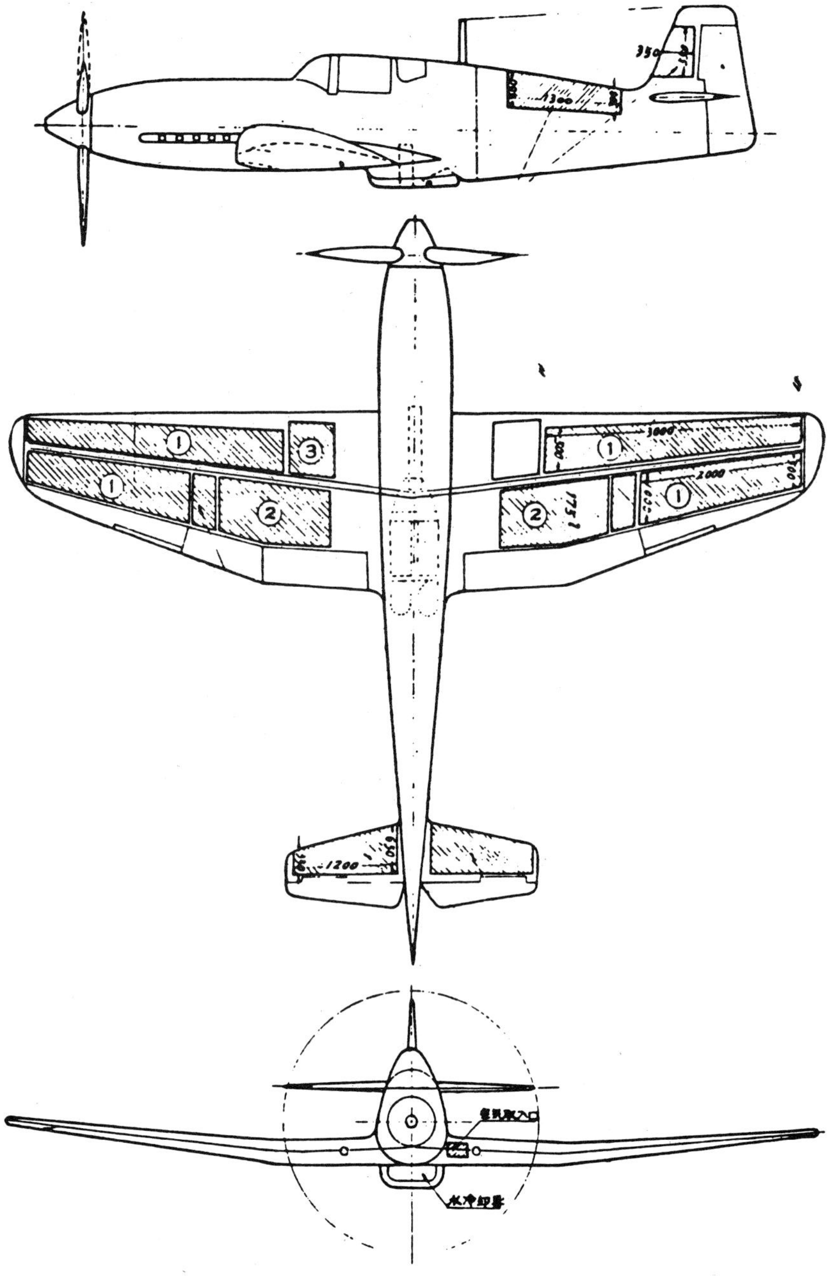

Project P 1078, Heinkel's attempt to bring the He 100 back to life.

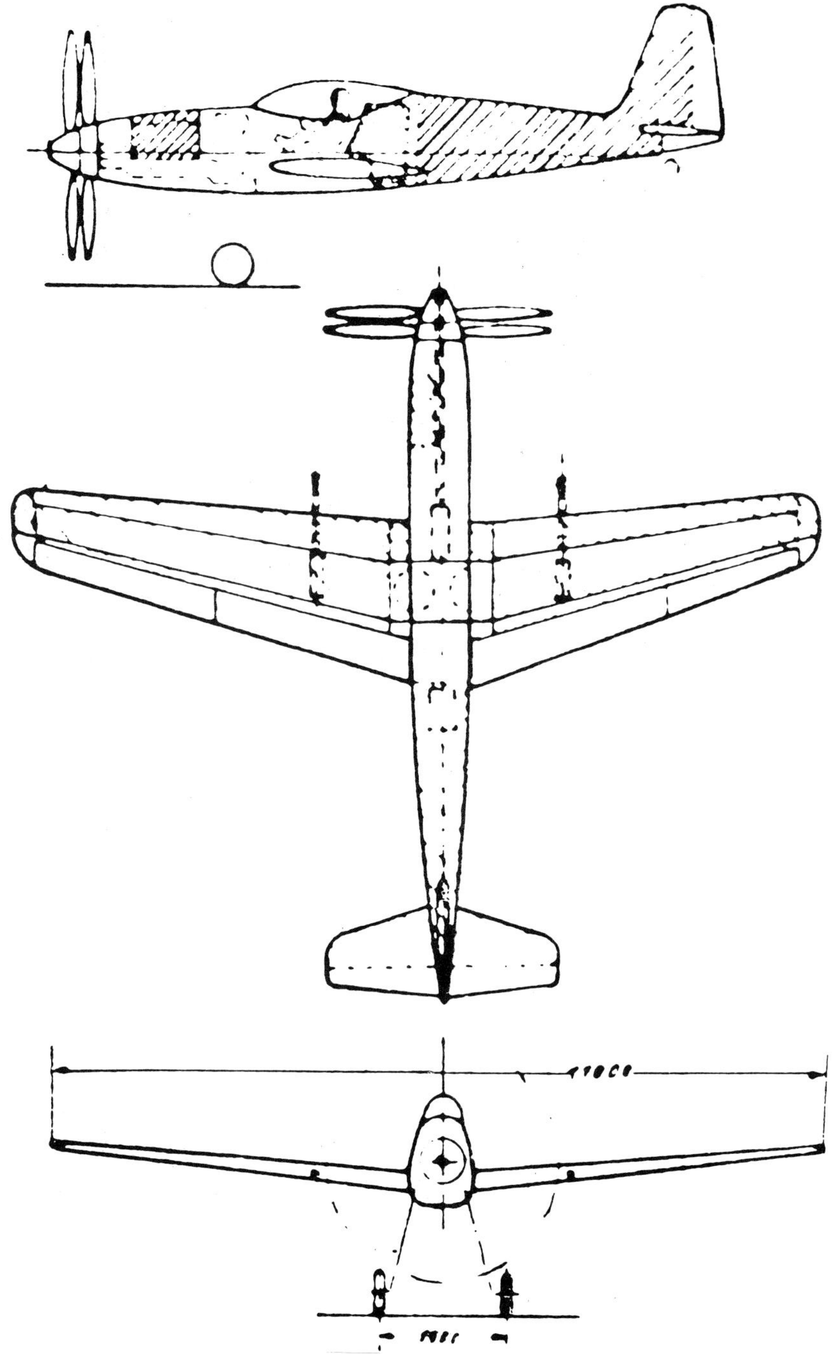

A Japanese version of the He 100, called the He 113.

He 100 V-7, factory number 3003

Registration D-IQRS, first flight February 1939, sent to Rechlin on June 9, 1939. The V-7 had a new heat exchanger in the oil system. High-speed clocked flights without armaments were made with it. This plane too was later sold to Russia.

He 100 V-8, factory number 1905

First flew on December 1, 1938 with production DB 601 Aa engine, first flight with DB 601 R record engine on January 8, 1939. Hans Dieterle set an absolute world speed record with this plane after Heinkel's chief pilot Gerhard Nitschke, after his "experiences" with this type, showed no great interest in gaining the laurels. After V-3 was wrecked, V-8 was prepared for the record run in the same way (short wingspan, flat windshield, special engine with short life span). This flight is a chapter in itself, too big to fit into this volume. Let us only say: The already described difficulties were never completely overcome, and in addition, the highly tuned motor required special racing spark plugs that had to be changed immediately before take-off, plus a number of other large and small modifications, all of which had to be dealt with — but Dieterle tried it — and won! With this comparatively delicate plane he flew 37.39 kph faster than Agello with his twin-engine powerhouse. In the German press, Dieterle and his Heinkel "112 U" were lauded ecstatically. For the record flight, the plane was not painted (to save weight) and showed a spotty exterior with its smoothed and spackled spots. Only after the flight was it given the usual paint job and D-IDGH registration. Later it was exhibited as a "normal" military plane with 42C+11 markings (merely pasted over the civilian registration — and it should have been 42+C11). Since V-8 was not sold to Russia, and since the plane displayed at the German Museum in Munich was not the original record plane, further record flights were forbidden after the Me 209's flight, and series production did not take place, one can suspect that the cell must have been kept for crash testing.

He 100 V-9, factory number 3004

Registration D-IWOP, originally equipped with weapons but used for crash testing shortly thereafter.

Above: The Japanese Kawasaki Ki 61 "Hien."

Right: The Russian LaGG3 (LaGG = Lawochkin, Gorbunov and Gudkov).

He 100 V-10, factory number 3005

Registration D-ITLR, originally planned for crash testing, then fitted with the short speed-record wings and flat cockpit canopy of V-8 and displayed at the German Museum in Munich as the "He 112 U record plane." It was destroyed in a bomb attack.

He 100 Exp.

Three planes were built especially for sale to Japan and were designated He 113 for the Japanese, but not noted in the RLM type list. These planes corresponded to the D type (see below). They were not copied in Japan, but their technical evaluation resulted in the Kawasaki Ki 61. In Russia too, the He 100 was not directly copied but studied very thoroughly; for example, the Soviet LaGG 3 fighter, as noted by German technicians, showed certain parallels to the He 100.

He 100 D

Various registrations for the total of ten planes built. When test models were built, Heinkel hoped that this model would be built in series. Every time they hoped to be able to sell this plane to the RLM, the model was given a series designation (A series, B series, etc.). Meanwhile, changes, improvements, simplifications and such were made feverishly, until the end of development was attained with the V-10, and this was called the D series, of which 25 planes were to be built (by Ernst Heinkel's own decision), but this was obviously interrupted. Meanwhile, surface cooling had been given up and normal cooling was reinstated, which caused the wingspan to be extended to 10.80 meters. With that plus weapon installation, the plane's speed advantage over the Bf 109 was much decreased. As far as is known, no planes were delivered directly to the troops. The D series was given the original Luftwaffe paint job and squadron markings (of non-existent units), but only served as the "new He 113 fighter" for propaganda, and was also registered by Heinkel for object protection. Whether it ever saw actual service or combat is not known.

The Me 209 (above), which took the He 100's world record 27 days later (some 9 kph faster), could not be remodeled as a "normal" fighter plane. After several rebuildings (below), the Me 209 looked very much like the Fw 190 D but had no advantage over the Bf 109. Its development was halted.

He P 1076

In 1944 the He 100 was resurrected again and Project 1076 was developed from it. This was a fighter with wings angled slightly frontward and two opposed three-bladed propellers. This high-speed fighter was not actually built, since the jet fighter had meanwhile been given precedence.

This short treatment has not told anywhere near everything about the He 100, especially as concerns the people and the associated events — which include many contradictions — but is certainly of use as a general overview. Supporting information, or contradictory data, will always be welcome.

TECHNICAL DATA

He 100	Wing-span meters	Length	Height meters	Weight kg* meters	Engine	HP	kph**	Range km	Ceiling	Armament
V-1	9.42	8/17	3.60	2158	DB601Aa	1050	620	980	10000	2 MG17, 1 MG/FFM planned
V-2	9.42	8.17	3/60	2470	DB601Aa	1000	622	1030	10000	2 MG17, 1 MG/FFM planned
V-3	7/60	8.18	3.60	2439	DB601Aa	1020	690	1100	10000	–
V-4	9.42	8/20	3/60	2540	DB601Aa	1020	670	1050	11000	2 MG17, 1 MG/FFM planned
V-5	9.42	8.20	3.60	2437	DB601M	1030	682	1120	11000	–
V-6	9.40	8.20	3.60	2580	DB600Aa	1030	690	1080	11000	–
V-7	9.40	8.20	3.60	2580	DB600Aa	1030	690	1080	11000	–
V-8	7.60	8.18	3.60	2475	DB601R	1800	770	–	–	–
V-9	9.40	8.20	3.60	2590	DB601M	1030	660	950	10000	2 MG17, 1 MG/FFM
V-10	9.40	8.20	3.60	2590	DB601M	1030	660	950	10000	–
He 100D	9.42/10.80	8.20	3.60	2500	DB601M	1030	670	1010	11000	2 MG151 or 2 MG17, 1 MG/FFM
P 1076	11.00	9.64	–	4380	DB603LM	1800	855	1600	14500	2 MK 108, 1 MK103***

* flying weight
** without height limitations
*** projected

Schiffer Military History

Schiffer Publishing Ltd.
1469 Morstein Rd.
West Chester, PA 19380
Phone: (215) 696-1001 FAX: (215) 344-9765

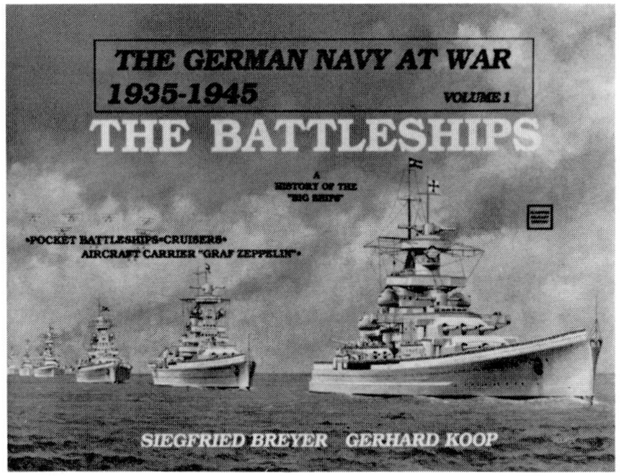

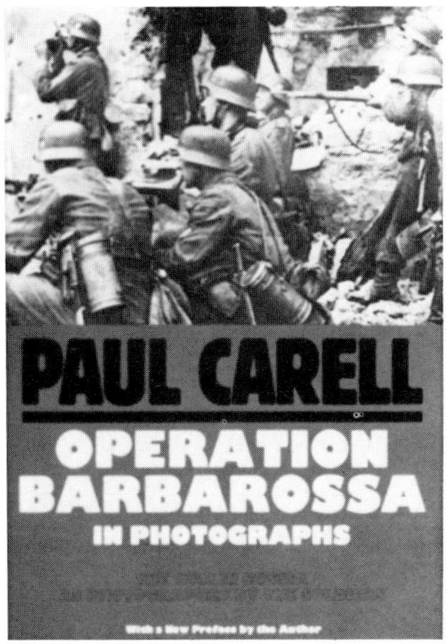